Robert Ochieng

A Review of degradation status of the Mau Forest and Possible Remedial Measures

GRIN Verlag

Bibliografische Information der Deutschen Nationalbibliothek:

Die Deutsche Bibliothek verzeichnet diese Publikation in der Deutschen National-
bibliografie; detaillierte bibliografische Daten sind im Internet über http://dnb.d-
nb.de/ abrufbar.

Imprint:

Copyright © 2009 GRIN Verlag GmbH
Druck und Bindung: Books on Demand GmbH, Norderstedt Germany
ISBN: 978-3-640-73759-8

This book at GRIN:

http://www.grin.com/en/e-book/156873/a-review-of-degradation-status-of-the-
mau-forest-and-possible-remedial

The Mau Forest Complex, Kenya: A Review of Degradation Status and Possible Remedial Measures

Robert M. Ochieng

Abstract

Deforestation and degradation of forests continue at alarmingly high rate, particularly in the tropics. Kenya's annual deforestation rate is estimated at 0.5%, putting at stake the survival of the timber industry and livelihood of forest dependent communities. The Mau forest is one of few remaining indigenous forests in Kenya with high deforestation rate. The forest supports the livelihood of the indigenous and surrounding communities and is major water catchment for the Eastern Africa region. This paper discusses the importance of the Mau forest and impacts of its degradation on the indigenous, national and regional communities, and proposes possible strategies to curb degradation of the forest. It is shown that degradation of the forest stems from activities of the surrounding communities, overpopulation and weaknesses in national laws and their enforcement. Several strategies are suggested; including involvement of the indigenous community in forest management, population control and the implementation of far reaching reforms in the forest and land sectors. It is recommended that since the benefits of the Mau forest are international, a debt-for-nature swap or similar schemes should be introduced to free national income for development and reduce the reliance on forest resources.

Keywords: *Deforestation, Tropical forests, Land use, Indigenous people, Ogiek, Forest communities, forest policy and law enforcement, solutions*

1.0.   Introduction

Deforestation and degradation of the world's forests are continuing at an alarmingly high rate (FAO, 2005). FAO (2005) estimates the current global annual deforestation rate at 13 million ha. Deforestation and degradation have been used interchangeably and defined variously. This discussion paper adopts the Food and Agricultural Organization of the United Nations (FAO) definitions which defines deforestation as the continuous conversion of forest lands to other land uses with significant reduction in tree cover and degradation as the loss of forest cover without noticeable change in canopy cover (FAO, 2003). High deforestation rates are particularly recorded in the tropics, where South America records the highest rate at 4.3 million ha per year followed closely by Africa at 4.0 million ha per year (FAO, 2005).

The implications of tropical deforestation are many and disturbing. Tropical forests contain over a half of the world's fauna and over 70% of the world's flora (Frey, 2002). These forests maintain ecological processes that are vital for the sustenance of life. They are home to several indigenous groups (Laurance, 1999) and are a source of a range of goods and services that are the lifeline of countless indigenous and forest dependent communities. Continued deforestation in the tropics is therefore a threat to biodiversity conservation, maintenance of vital ecological processes and survival of indigenous communities. In the context of the current global environmental challenge, climate change, the most important function of tropical forests is perhaps their role as sources and sinks of carbon. Continued deforestation and degradation of tropical forests is estimated to account for about 20% of the total anthropogenic green house gas emissions (IPCC, 2007). There is, therefore, urgent need to address deforestation and degradation of these forests.

With a forest cover of less than 2% of the total land mass (FAO, 2003; FAO, 2005), Kenya is considered one the countries with the lowest forest cover in Africa, and where substantial degradation occur in all the major forest blocks. Between 1990 and 2000, Kenya's forest cover reduced by 93 000 ha per year (FAO, 2003), translating to an annual rate of 0.5%, which remains constant to date (see FAO, 2005). While this figure is lower than Africa's average of 0.62%, it is significantly higher than the world average of 0.18% and also considering the country's total forest cover. Agricultural expansion and increased population growth are the two critical factors underpinning deforestation in the country (Wass, 1995). With 80% of the Kenyan population dependent on agriculture (Lambrechts et al., 2005), there is increasing pressure to release more land for agriculture and hence the continued forest clearance.

The Mau forest complex is one of the few remaining indigenous forest blocks in Kenya. The forest covers an area of 400 000 ha (Lambrechts et al, 2005; UNEP, 2008), comprising both indigenous and plantation forests. It supports the livelihood of the surrounding agro-based communities by supplying wood to the tea factories and a host of other goods and services (Wass, 1995). The forest is also home to East Africa's largest forest dwelling community, the 'Ogiek' or 'Dorobo'. Increased population and the consequent demand for more land have led to massive degradation of the forest (Obare and Wangwe, n.d; Wass, 1995), thereby threatening the survival of the 'Ogiek' and surrounding communities. Degradation of the forest has further been exacerbated by unsustainable commercial exploitation and forest excisions (Kunga, 2003). The Kenyan government has, over the last three decades, made several attempts to curb further degradation of the forest. These have ranged from coercive methods such as eviction of forest dwellers, to creation of a boundary of tea plantations around the forest to buffer the forest from further encroachment (KFMP, 1994; Kagombe and Gitonga, 2006). However, none of these strategies have achieved any tangible results. Instead, they have resulted in the current predicament of the 'Ogiek' and other forest surrounding communities and have worsened degradation of the forest.

The aims of this discussion paper are three-fold. (1) To highlight the importance of the Mau forest and impacts of its continued degradation on the local, national and regional communities. (2) To highlight the plight of the indigenous forest community (Ogiek) and the surrounding population, and (3) to propose possible strategies through which the Mau forest can be restored and conserved. The paper begins with a brief review of forest development in Kenya, highlighting the importance of the sector to the economy and wellbeing of the Kenyan people. The second section describes the Mau forests and its contribution to local livelihoods and the implications of its continued degradation on the national and regional economies. This section also discusses the plight of the traditional forest dwelling community, the 'Ogiek'. The next section summarizes previous efforts to curb deforestation in the forest and their impacts on forest dependent communities. The last section presents possible strategies to mitigate degradation of the forest, and ends with the conclusion.

2.0. Forestry Development in Kenya and Importance of the Forest Sector

The history of industrial forestry in Kenya can be traced back to the early 1900s, during the construction of the Kenya-Uganda railway. At the time, the colonial government was concerned over the sustainability of wood supply for the then steam-engine driven trains (KFMP, 1994). A decision was thus made to establish plantations to supply industrial wood. Noting that indigenous species could not meet the country's growing wood demand because of their slow growth, it was decided plantations of fast growing exotic species be established (www.kfs.co.ke). Industrial forestry therefore started with this ambitious programme and by the late 1970s the area under plantations had reached over 160 000 ha (Odera, 2001). These comprised mainly of cypress (*Cupressus lusitanica*), pines (mainly *Pinus patula* but also *P. radiata*) and Eucalyptus species. The success in plantation establishment was made possible by the use of the 'Shamba' system (Kagombe and Gitonga, 2006). This was a modified form of the 'Taungya' system in which farmers were allocated forest land to grow food crops alongside tree crops. They cared for the tree crops for two to three years before they were allocated new plots after successful establishment of the tree crop.

The establishment of plantations relieved pressure on the indigenous forests. The management objective for these forests changed to preservation with extractive uses being abolished except for collection of fuel wood and non-wood forests for non-commercial purpose by surrounding communities (Kagombe and Gitonga, 2006). The period leading to the 1980s marked the blossoming time for forestry in Kenya. Production of food crops alongside trees ensured food security (KFMP, 1994). Sustainability of wood supply from plantations led to the establishment of a vibrant timber industry and allied enterprises and the mushrooming of forest towns such as Molo, Elbagon and Maji Mazuri in the Rift Valley. These provided employment to a number of people and improved the livelihood of many households, especially around the major forest blocks.

Kenya's forests play a significant role in the livelihood of the people and the country's economy. Estimates indicates that about 530 000 households, representing some 2.9 million people, live within five kilometers of the forest (Odera, 2001; Lambrechts et al., 2005; Wanyiri, 2007). Lambrechts et al. (2005) observe that these households are directly reliant on the forest as a source of livelihood. About 80% of rural households and 70% of urban dwellers are dependent on wood for heating and cooking (Senelwa, 2005). The forest sector provide direct employment to 50 000 people and indirect employment to 300 000 people (Wanyiri, 2007; Lambrechts et al., 2005). Forestry contributes 3% of the Gross Domestic Product and 13% to the informal economy (Wanyiri, 2007; Lambrechts et al., 2005). They provide a range of non-timber forest goods such as carvings, gums, resins and honey. Oduor et al. (2002) and KAFU (2000) estimates the annual turnover of exports of non-timber forest products at over 30 million Euros. Kenya's forests also provide ecological services such as biodiversity conservation, stream-flow regulation and soil stabilization.

In spite of the contribution of forests to the wellbeing of Kenyans, the sector has, since the 1980s, been facing several challenges. The forest cover has constantly been dwindling. Today, Kenya's forest cover is estimated at less than 2% of the land area (FAO, 2003; FAO, 2005; Mbugua, 2008), down from a cover of over 3% in the early 1980s. Plantations which sufficiently supplied the country's wood demand (Oduori and Ogweno, 2001) have today reduced to only about 120 000 ha (Mbugua, 2005). A wood supply deficit of up to one million cubic meters has been forecasted beginning the year 2015 (Ototo, 2001; Oduori and Ogweno, 2001), putting at stake the livelihood of the over 2.9 million people directly dependent on the forest. This scenario has put to the fore questions on where Kenya's forestry lost track.

2.1. The Mau Forest and the Predicament of the 'Ogiek' Community

2.1.1. The Mau Forest

The Mau forest was gazetted as a forest reserve by the colonial government in 1932 (Obare and Wangwe, n.d.; Kunga, 2003). The forest is located on the Mau escarpment in the Great Rift Valley. It straddles Kericho, Bomet, Nakuru, and Narok districts. Kericho and Bomet districts also host most of the tea plantations in the country. Nakuru district hosts the Lake Naruku National Park which is known globally for its richness in flamingoes and is a major tourist destination. Narok district is home to the Maasai Mara game reserve, where the historic wildebeest migration is a major tourist attraction. The wildebeest migration was in 2006 named the Seventh Wonder of the World by the United State's ABC television channel. Tourism and tea production are today Kenya's leading income sources and top foreign exchange earners, with a combined turnover in excess of 200 million Euros (UNEP, 2008).

The Mau forest is a major water reservoir and forms the catchment of Rivers Sondu, Mara and Ewaso Ngiro, and nine other small rivers (Obare and Wangwe, n.d). River Sondu drains into

Lake Victoria, which straddles the three East African states, Kenya, Uganda and Tanzania and is the source of River Nile. River Mara drains into Lakes Bogoria, Naivasha and Nakuru, while River Ewaso Ngiro drains into L. Natron in Tanzania (Obare and Wangwe, n.d). UNEP (2008) estimates the annual direct and indirect revenues from the Maasai Mara and L. Nakuru national parks at 50 million Euros. Continued supply of water from the forest is vital for the survival of flamingoes in L. Nakuru and wildlife in the Maasai Mara, and the Serengeti in Tanzania. Estimates by KFMP (1994) put the catchment protection value of the forest at over eight million Euros per year. The hydropower potential of rivers flowing from the Mau is estimated at 535 MW, with an annual value of 80 million Euros (UNEP, 2008). Power remains one of the scarcities in the country and is seen by many economists as a major hindrance to the realization of Vision2030. Exploitation of these hydropower resources would not only stem the current perennial power shortages, but also reduce Kenya's dependence on imports of electrical energy from Uganda. However, this can only be possible when the rivers' catchment is conserved.

The Mau complex is a habitat for mammals of international conservation concerns such as the Yellow backed Duiker, the Bongo, the Golden Cat, leopards and elephants (Obare and Wangwe, n.d). It contains some of the rarest, threatened and endangered avian species in East Africa (Birdlife, 2008). The forest is rich in distinct species assemblages, courtesy of its unique location on the Mau escarpment that permit altitudinal and microclimatic variations. With over three million people within the Eastern African region and beyond estimated to rely on goods and services from the Mau (Lambrechts et al., 2005), continued deforestation of the forest is a threat to the whole international community. This perhaps explains the wide coverage of the Mau crisis by both national and international media such as the Reuters.

2.1.2. The Indigenous Forest Community (the 'Ogiek') and other forest-dependent communities

The Mau Complex is home to the only remaining forest-dwelling community in East Africa, the 'Ogiek' (Wass, 1995). This group of people has since time immemorial lived and depended on the forest for subsistence and shelter (KFMP, 1994; Wass, 1995; Obare and Wangwe, n.d). The traditional way of life of this hunter-gatherer community consisted of collecting wild fruits and nuts, hunting and honey harvesting (Kunga, 2003). The community had a well organized management system for the forest (Kunga, 2003; Obare and Wangwe, n.d). They divided the forest into blocks among their clans, which in turn divided the blocks among their family lineages. Under the watchful eye of the community elders, each family was expected to ensure conservation of the forest block under its jurisdiction. Since the community was not engaged in any form of farming, there were no cases of forest clearing. Forest fire incidences were rare since only experienced men were allowed to harvest honey (Kagombe and Gitonga, 2006; Kunga, 2003). Quotas were imposed on harvesting of sensitive tree species such as *Olea uropea*, *O. hochstetteri* and *Dobeya goetzeni* (Obare and Wangwe, n.d.). In this way, the community ensured sound conservation of the forest.

However, the harmonious relationship between the community and the forest started experiencing problems with the gazettement of the forest in 1932. The established Forest Act rendered them illegal occupants of the forest (Kagombe and Gitonga, 2006). The then colonial government made several attempts to evict them but without success (Kunga, 2003; Obare and Wangwe, n.d). This move, instead, forced them deep into the forest. The evictions continued into the post-colonial Kenya, with the worst evictions occurring in the periods leading to 1992 and 1997 general elections. During these periods, the 'Ogiek' were moved out of the forest with the promise that they would be resettled elsewhere. This left them to settle on the forest fringe and around forest stations, while a good number of them moved back into the forest barely after a couple of months (Sang, 2002; Kunga, 2003). When time for resettlement came, very few of them, if any were allocated land (Kagombe and Gitonga, 2006; Sang, 2002). The 'Ogiek', through their umbrella organization (The Ogiek Welfare Association) claim that areas of the forest excised for resettlements were mainly allocated to people from other communities and well connected civil servants and military officers (Kunga, 2003; Lambrechts et al., 2005; Sang, 2002). This prompted them to move to court in 1997 to stop any further eviction (Sang, 2002). The suit has to date, however, not been determined.

As a result of these developments, the 'Ogiek' are today among the poorest communities in Kenya. Their lives have been shattered and their traditional way of life is no longer possible.

According to Amnesty International (2007) the evictions were executed with excessive force, accompanied with burning of houses, schools and crops. This has left the people living in deplorable conditions. Most of their children were forced out of school (Kunga, 2003). This perhaps explains why the 'Ogiek' have one of the highest illiteracy levels in Kenya. The overall impact of this has been an increase in crime rate and high prostitution levels in the major areas where they live such as Molo and Elburgon.

2.2.   Previous and Ongoing Efforts to Curb Deforestation of the Mau

The Kenyan government has implemented a number of strategies to curb further degradation of the Mau forest over the years. Key among these strategies include, evictions of the forest dwelling community; establishment of a boundary of tea plantations around the forest and plantations of exotic species; and imposition of logging bans on both indigenous and plantation forests. The eviction of people has mostly had negative results. Most importantly, it has led to landlessness and escalated the predicament of the indigenous and forest dependent communities. Increased landlessness forced the government to excise some parts of the Mau forest for settlement. Lambrechts et al. (2005) estimates that between 1967 and 1989, the government excised 24 ha of Southwest and 300 ha of Eastern Mau for settlements. In 2001, a total of 67 000 ha of the Mau forest were excised for settlements (UNEP, 2008). These excisions have been done in unsystematic manner, resulting in fragmentation of the forest. Several studies (e.g. Saunders et al., 1991; Collinge, 1996) have shown that fragmentation of forests decreases forest core habitats that are vital for maintenance of biodiversity and microclimates for productive agriculture, a scenario detrimental for a country heavily dependent on agriculture such as Kenya.

In the 1980s, in response to forest encroachment, the government enacted a policy to enable the establishment of boundary of tea plantations around the major forest blocks, including the Mau. This was restricted by law to within 100 meters of the forest edge (KFMP, 1994), and was carried out in all the major forest blocks. Because of lack of strict law enforcement the system was abused by unscrupulous officers who hived off much land than was allowed. In the Mau, some of the areas that were clear felled for this purpose were never put under tea but were converted to agricultural land (KFMP, 1994; Kunga, 2003). Experts assert that through this scheme, about 28% of Mau forest was hived off and that much of the land that was cleared for this purpose have today been converted to tea plantations and private ranches (UNEP, 2008). This shows that the plan did not achieve its intended purpose.

Attempts have also been made to reforest and restore parts of the Mau that have been severely degraded with plantations of exotic species. As early as 1943, about 10% of the Mau forests had been cleared to pave way for plantations (Obare and Wangwe, n.d.). However, while plantations meet timber needs, Noss and Cooperrider (1994) showed that they do not necessarily restore the original species diversity and ecological functioning of the forest, and that some exotic species may be incompatible with biodiversity conservation. Because of this, Pimentel et al. (2000) and Ricketts et al. (2004) advocate for use of indigenous species in forest restoration initiatives. However, this measure has not been tried on a significant scale in the Mau, except for disjointed small scale monocultures of a few indigenous species such as *Juniperus procera*, *Prunus Africana*, and *Podocarpus* species and disparate enrichment plantings in scattered parts of the forest.

Logging bans have been imposed to curb illegal logging in all forest blocks. The first ban was imposed in 1986 to prohibit logging in indigenous forests (see Odera, 2001). However, even with the imposition of the ban, cases of forest degradation and illegal logging remained rampant. Continued widespread illegal activities in state forests prompted the government to impose a ban on plantations by a presidential degree in 2000 (Kagombe et al., 2005). While the impacts of the ban on the Mau forest is not clear, indicators depict that it has had severe effects on the timber industry and allied enterprises, and the livelihood of forest dependent communities. For instance, Kogombe et al. (2005) estimate that 400 saw mills and allied enterprises in the Rift Valley closed down as a result of the ban, rendering some 150 000 people jobless. Furthermore, prices of wood products have shot up due to reduced supply (Kagombe et al., 2005; The Dailly Nation, Tuesday 30th May 2006), thereby providing greater incentive for illegal logging. Studies on impacts of logging bans in the Asia-Pacific region (e.g. Fao, 2000;

Macan-Marker, 2001; Waggener, 2005; Sureeratna, 2005) have indicated that logging bans are an ineffective strategy to curb deforestation and forest degradation.

2.3. The Way Forward

From the preceding discussion, it is clear that most of the problems facilitating degradation of the Mau forest stems from activities of surrounding populations, population increase and weaknesses in national laws and policies. Measures aimed at addressing degradation of the forest must, therefore, specifically address these key issues. Below are possible solutions to the degradation of the Mau forest. The solutions build on weaknesses of previous measures to address forest degradation in Kenya and borrow from experiences in other regions of the world, particularly the South American and Asian regions.

2.3.1. Community involvement

Although the 'Ogiek' have lived in the Mau for centuries, today their actions are in direct conflict with forest management policy. However, it is important to recognize that successful forest conservation programs are those that have been done in cooperation with the traditional forest owners (Laurance, 1999). The 'Ogiek' require adequate land for their culture and way of life to survive, and their strong case for cultural sustenance is strengthened by international laws on indigenous rights. However, there are currently no direct provision in the Kenyan Constitution to safeguard indigenous culture and land. The government should involve the community in managing the forest. This should not be impossible as the new Forest Act (GOK, 2005) allows joint management of forest between communities and the forest authority under Community Forest Associations. There is urgent need to allocate the community a part of the forest where they can exercise their culture while also safeguarding the forest.

There are also immigrant communities who have invaded the forest in the hope that they could be settled along with the 'Ogiek' (see KFMP, 1994, Wass, 1995; Kunga, 2003). These people lack sufficient income and resources to buy inputs for sustainable agriculture. As a result, they use unsustainable farming practices that further degrade the forest. Furthermore, their dependence on the forest for subsistence and commercial activities such as charcoal burning and firewood collection has escalated degradation of the Mau. While majority of them may not be genuine in their claim of landlessness, the Ndungu Land Report can be used as a basis for identifying genuine landless people and those who have deliberately invaded the forest. The government should purchase land elsewhere and settle those who are genuinely landless. Along with this, the government through the ministry of agriculture should educate them on farming methods that are not only compatible with conservation but also profitable. This will not only reduce forest degradation, but will also improve income and livelihood and hence ensure economic growth.

2.3.2. *Population control*

It is almost universally agreed that population pressure in the most crucial underlying cause of deforestation all over the world (Laurance, 1999; FAO, 2005), and it is no doubt that it is a key driver of forest degradation in Kenya (e.g. KFMP, 1994; Glenday, 2006). Over the last two decades Kenya's population has doubled, with annual growth rate rising to 2.4%. This increase in population puts more strain on the already dwindling natural resources. The population around the Mau forest has doubled over the years, with the impact being demand for more agricultural land. While issues like population control are sensitive in countries like Kenya, some forms of population control measures have been implemented in the past through the Family Planning Program. Perhaps the government could first educate the people on the need to have smaller households and then introduce policies on the maximum family size. Although this suggestion may look abstract and superficial, it has effectively been applied in China (Hou Zheng-Yang, pers. Comm.).

2.3.3. *Effective law enforcement and coordination between various arms of the government*

Enforcement of forest and environmental laws has been wanting over the last two decades. This is mainly due to the massive retrenchment in the 1990s in response to demands by international development banks that greatly affected the forest sector (Obare and Wangwe, n.d.). This reduced the number of for instance forest guards to the extent that today there are barely two guards for every management unit of the Mau forest, spanning several hundred hectares. These officers also lack the necessary equipment for successful accomplishment of their duties. Too much power vested in the executive that gives the police more power in handling forest offence, gaggles the operations of the forest guards and provides a leeway for forest offenders due to the corrupt nature of the force (see Sang, 2002; Kunga, 2003). This problem is exacerbated by low fines meted on forest offenders which have acted more as incentives for illegal activities in the forest. While the revamping of forest management through the enactment of the new Forest Act and the establishment of the semi-autonomous Kenya Forest Service is promising, there is need to speed up reforms in the sector. The civil society and environmental groups should put more pressure on the government to force it to ensure that logging regulations and other environmental laws are applied to the letter.

The failure of resettlement efforts have been blamed on the lack of coordination between the key government ministries involved (Sang, 2002), namely the ministries of Forestry (previously Ministry of Environment and Natural Resources), Provincial Administration, Water and Land. The actions of these ministries have in many situations been contradictory, thereby rendering them counterproductive. Schomberg (1999) (cited in Laurance, 1999) argues that there is no point in one Minister (e.g. for Forestry) cracking down on encroachers while the other (e.g. for Land) is settling people right in the middle of the forest. Reinforcing this weakness in coordination is the lack of clear policies on eviction and resettlement of people (Amnesty International, 2007). There should therefore be a clear coordination between the ministries in resettling people and tackling deforestation in the Mau. The government should formulate and enact policies on evictions and resettlement of people as recommended by UNESCO and Amnesty International Kenya, which can withstand legal challenges if eviction and resettlement programs are to succeed. The civil society and local and international environmental non-governmental organizations can be tapped for this purpose.

2.3.4. Debt-for-nature swaps

Land is a particularly emotive issue in Kenya. If the civil unrest triggered by the disputed presidential election in 2007 is anything to go by, then inequalities in land ownership is one big problem in Kenya. There is need to implement comprehensive land reforms in the country. The report of the commission of inquiry into land issues, or Ndungu Report as it is famously known, provides a good background to start from. Successful implementation of land reforms, however, depends almost exclusively on political will. For instance, attempts to address the Mau crisis by the executive have been thwarted by stiff opposition from Rift Valley legislators. Two things can be done to address this problem. While the Kenyan government does not rely so much on foreign aid, it is one the major debt-burdened countries. The international community can introduce what conservationists refer to as 'debt-for-nature swaps' (for a review of debt-for-nature swaps see Kahn and McDonald, 1995) as a means of securing forest conservation in Kenya.

The relationship between deforestation and environmental degradation in the third world and dept are well documented. In a study of the role of dept relief and tropical deforestation, Kahn and McDonald (1995) established that reducing third-world country debt reduces deforestation. Their argument that offering debt relief for Third-World countries in exchange for environmental protection can help reduce tropical deforestation is supported by other studies (e.g. Van Soest and Lensink, 2000). Debt-for-nature swap is a scheme in which deforesting countries are given debt relief in exchange for forest protection and have been implemented widely throughout the globe. The international community and conservation organization can negotiate an agreement with the Kenyan government in which Kenya is offered a debt relief conditioned on successful conservation of its forests. To ensure the agreement is not barred by aspects of sovereignty, the agreement is designed in such a manner that the relief is also negatively correlated with forest destruction as has been suggested by van Soest and Lensink (2000). Since the agreement is international, it becomes obligatory and with then not be subjected to opposition by regional groupings of legislators as is the case today with the Mau crisis.

3.0.   Conclusion

This discussion paper has reviewed the status of degradation of the Mau forest. It is apparent that degradation of the Mau continues at the peril of the local, national and international community. In particular, degradation of the forest threatens the livelihood of the Ogiek and surrounding communities and the survival of tourism and tea industry, which are key pillars of the Kenyan economy. It has been shown how weaknesses in national policies, increased population and unsustainable practices by surrounding communities have impacted the forest. Possible solutions to the Mau crisis have been suggested. Key among them is the need to involve indigenous and surrounding communities in the management of the forest. This should be accompanied an effective population control scheme to address the problem of overpopulation. As weaknesses in laws and their enforcement are a key factor of degradation, the government should ensure forestry and environmental laws are effectively adhered to. This should be accompanied by far reaching reforms in the forest and land sectors, and close coordination among government ministries in tackling degradation of the Mau. It is noted that Kenya will need to use its natural resources to meet it social and economic need for development purposes. We suggest debt-for-nature swaps to free more national resources for development activities and reduce reliance on natural resources. Emerging conservation schemes such as Reduced Emissions from Degradation and Deforestation (REDD) and Voluntary Carbon Markets (VCM) are schemes that could be exploited to curb deforestation in the Mau forest.

4.0.   References

Amnesty International. 2007. Nowhere to Go: Forced Evictions in Mau Forest, Kenya. Briefing
       Paper, May 2007. Available:
       http://www.amnesty.org/en/library/asset/AFR32/006/2007/en/dom-
       AFR320062007en.pdf (Accessed 19/02/2009).

BirdLife International. 2008. BirdLife's online World Bird Database: the site for bird
       conservation. Version 2.1. Cambridge, UK: BirdLife International. Available:
       http://www.birdlife.org (accessed 18/2/2009).

Collinge, S. K. 1996. Ecological consequences of habitat fragmentation: implications for
       landscape architecture and planning. Landscape and Urban Planning 36: 59 – 77.

FAO. 2005. State of the World's Forests. Food and Agricultural Organization of the United
       Nations (FAO), Rome, Italy.

FAO. 2003. State of the World's Forests. Food and Agricultural Organization of the United Nations (FAO), Rome, Italy.

FAO. 2000. Logging Bans have Mixed Successes in Conserving Forests. Food and Agriculture Organization of the United Nations. Available: www.fao.org/news/2000/000502-e (Accessed 17/06/2006).

Frey, E. F. 2002. Tropical Deforestation in the Amazon: An Economic Analysis of Rondania, Brazil. Political Economy 11 (2002).

Glenday, J. 2006. Carbon Storage and Emissions Offset Potential in East African Tropical Rainforests. Forest Ecology and Management 235 (2006): 78 – 83.

IPCC. 2007. Climate Change 2007: The Physical Science Basis: Summary for Policymaker. Available: http://www.ipcc.ch/pdf/assessment-report/ar4/wg1/ar4-wg1-spm.pdf (Accessed 27/02/2009).

KAFU. 2000. What are the non-timber forest products. An article in KAFU News Update, a newsletter for the Kenya Association of Forest Users, Issue No. 1 of May 2000.

Kagombe, J. K. and Gitonga, J. 2005. Plantation Establishment in Kenya: A Case study of the Shamba System. Available: http://www.kenyaforestservice.org/pub/NRC%20review%202005%20case%20studies%2011th%20May.pdf (Accessed 24/02/2009)

Kagombe, J. K., Gitonga, J., Gachanja, M. 2005. Management, socioeconomic impacts and implications of the ban on timber harvesting. Policy Brief No. 1. Kenya Forest Working Group (KFWG). Pg. 4.

Kahn, J. R. and McDonald, J. A. 1995. Third-World Dept and Tropical Deforestation. Ecological Economics 12 (1995): 107 – 123.

KFMP. 1994. Kenya Forestry Master Plan. Forest Department, Ministry of Environment and Natural Resources. Government Printers, Nairobi, Kenya.

Kunga, N. 2003. Challenges in Forestry Conservation in East Africa. Is Community-Based forestry the Key to Forest Survival? East African Ecotourism Development and Conservation Consultants. Nairobi, Kenya (Unpublished Report).

Lambrechts, C., Nkako, F. M., Gachanja, M. and Woodley, B. 2005. Mau Forest Status Report 2005. Available: http://new.unep.org/dewa/assessments/EcoSystems/Land/mountain/pdf/maasai_mau _report.pdf (Accessed 24/02/2009).

Laurance, W. F. 1999. Reflections on the Tropical Deforestation Crisis. Biological Conservation 91 (1999): 109 – 117.

Macan-Markar Marwaan. 2001. Logging Bans in Asia Prove Ineffective. Available: www.earthisland.org/borneo/news/articles/011112article.html (Accessed 17/06/2006).

Mbugua, D. K. 2005. Forest production system in Kenya. In: Muchiri, M. N., Kamondo, B. Tuwei, P., Wanjiku, J. (eds) 2005. Recent Advances in Forestry Research and Technology Development for Sustainable Forest Management. Proceedings of the 2nd KEFRI Scientific Conference, Muguga, Kenya, 1st - 4th 2004.

Noss, R. F., Cooperrider, A. 1994. Saving Nature's Legacy: Protecting and Restoring Biodiversity. Defenders of Wildlife and Island Press, Washington, DC.

Obare, L. and Wangwe, J. B. n.d. Underlying Causes of Deforestation and Forest Degradation in Kenya. Available: http://www.wrm.org.uy/deforestation/Africa/Kenya.html (Accessed 22/02/2009)

Odera, J. A. 2001. Progress in Implementation of C & I for Assessing Progress Towards Sustainable Forest Management within the Dry-zone Africa Process. A Paper Presented to a Country-led initiative in Support of the United Nations Forum on Forests International Expert Meeting on Monitoring, Assessment and Reporting on the Progress toward Sustainable Forest Management. Yokohama, Japan. 5th - 8th November 2001.

Oduor, N., Kiunga, K. and Mbiru, S. 2002. Status of the Non-Wood Forest Products in Kenya. A Summary Report for the Technical Liaison Committee of the Kenya Forestry Research Institute (KEFRI) and the Forest Department.

Odwori, P. O., Ogweno, D. C. O. 2001. Modeling Forest Plantation Development in Kenya. Discovery and Innovation. (Special edition) 2001: 37- 48.

Ototo, G. O. 2001. Feasibility of integrated sawlog and pulpwood harvesting of softwood plantations in Western Kenya. M.Phil thesis, Moi University, Eldoret, Kenya.

Pimentel, D., Lach, L., Zuniga, R. and Morrison, Dong. 2000. Environmental and Economic Costs of Nonindigenous Species in the United States. BioScience 50 (2000): 53 – 65.

Ricketts, T. H., Hoekstra, J. M., Boucher, T. M. and Roberts, C. 2004. Confronting a Biome Crisis: Global Disparities of Habitat loss and Protection. Ecology Letters 8 (2004): 23 – 29.

Sang, J. K. 2002. The Ogiek Land Question. Paper submitted at the Indigenous Rights in the Commonwealth Project Africa Regional Expert Meeting. Cape Town, South Africa. 16th – 18th October 2002.

Saunders, D. A., Hobbs, R. J. And Margules, C. R. 1991. Biological Consequences of Ecosystem Fragmentation: A Review. Conservation Biology 5 (1991): 18 – 32.

Schomberg, W., 1999. Brazil Suspends issuing of Amazon Clearing Permits. Reuters News Service, 12 February 1999.

Senelwa, A. K. 2005. Forest products and services. In: Muchiri, M. N., Kamondo, B. Tuwei, P., Wanjiku, J. (eds) 2005. Recent Advances in Forestry Research and Technology Development for Sustainable Forest Management. Proceedings of the 2nd KEFRI Scientific Conference, Muguga, Kenya. 1st - 4th August 2004.

Sureeratna Lakanavichia. 2005. Impacts and Effectiveness of Logging Bans in Natural Forests: Thailand. Available: www.fao.org/docrep/0003/x6967e/x6967e09.htm (Accessed 17/06/2006)

UNEP. 2008. Mau Complex under Siege: Threats and Values. Available:
http://www.unep.org/dewa/assessments/EcoSystems/Land/mountain/pdf/Mau_crisis_
2008.pdf (Accessed 25/02/2009).

Waggener T. R. 2003. Logging Bans in Asia and Pacific: An Overview. The International Forestry
Review Vol. 15(3), September 2003. Common Wealth Forestry Association. Oxfordshire
OX26 6ZJ. UK. Available: www.fao.org/docrep/003/x6967e/x6967e04 (Accessed
17/06/2006)

Wanyiri, J. M. 2007. Forest Policy, Legal and Institutional Framework Information Sheet. Kenya
Forest Service, Nairobi, Kenya.

Wass, P. 1995. Kenya's Indigenous Forest Status. Management and Conservation. IUCN, Gland,
Switzaland and Cambridge, UK.